BEEKEEPING SIMPLIFIED WITH THE DRAYTON HIVE
INCLUDING PLANS FOR HOME CONSTRUCTION

Andrew Bax

ISBN: 978-1-914934-49-0

Published by Northern Bee Books 2022
Northern Bee Books, Scout Bottom Farm, Mytholmroyd,
Hebden Bridge HX7 5JS (UK).
www.northernbeebooks.co.uk
+44 (0) 1422 882751.

Book design by www.SiPat.co.uk

All rights reserved. No part of this publication may be reproduced, stored or transmitted in any form or by any means electronically or mechanically, by photocopying, recording, scanning or otherwise, without the permission of the copyright owners. © Northern Bee Books

Beekeeping Simplified with the
Drayton Hive
Including Plans for Home Construction

by Andrew Bax

Dedicated to my Susan
who has put up with a lot over the decades
but she does like honey

Contents

Preface	6
Introduction	8
A simplified approach to beekeeping	10
Essential features of the Drayton Hive	**12**
Horizontal format	12
All round, all-year insulation	12
Raised, small entrance	13
Observation window	15
Roof	16
Floor	17
Frames	18
Queen excluder	20
Division board	21
Cover cloth	21
Feeding	**24**
Harvesting	**26**
Honey	26
Beeswax	29

Hive hygiene	**30**
Seasonal management	**31**
Winter	**31**
Spring	**31**
Summer	**32**
Autumn	**33**
Some personal reflections on beekeeping practice	**34**
Uniting colonies	**36**
A case study	**37**
Who is the Drayton Hive for?	**39**
Further information	**41**
Home construction	**43**
Plans	**44**

Preface

This little book is really a tribute to Carass, a remarkable man who came to my rescue when, unexpectedly, I became a beekeeper.

A friend of mine had a couple of WBCs in his garden. One day he had hurried back from his office to undertake a routine manipulation in cold and blustery weather. He was under pressure of time and didn't bother to protect himself adequately. The bees resented his intrusion, and he was now in hospital. His wife pleaded with me to take the hives away.

I didn't know anything about bees or beekeeping, but I did have a little orchard. Luckily, I was introduced to Carass and we met the following evening. There had been a change in the weather, all was calm, and the bees were busily going about their business without any sign of the previous day's defensiveness. Carass turned out to be well into his eighties and crippled with arthritis. As he hobbled up to the hives on crutches, I began to wonder how much help he could be.

I need not have worried. Protected only by a moth-eaten veil, he gently lifted the roof from the first hive, breathing in the sweet, warm aroma. 'Ah,' he said, 'they've been on the sycamore today.'

From that moment I was hooked. I had no idea that some people could tell where bees have been from the smell of the hive; it was the first of many astonishing facts I was to learn. We waited until the evening got cooler and foraging bees had returned to their hives. In the meantime, Carass told me something about himself in his soft, Yorkshire accent.

In the 1930s he had a smallholding in the Dales, and 100 beehives in the hills. His only form of transport at the time was a pony and trap. To me it seemed impossibly long ago and a world away. However, he rose to some eminence in the echelons of beekeeping; he was much in demand as a judge at honey shows and, during the Second World War, he was appointed by the Ministry of Agriculture to bolster honey production. Now, instead of beekeeping, he spent his days making and repairing violins.

For the remaining few years of his life, he came to visit my hives whenever he could. Together we raised queens, studied pollen through his microscope and made polish from beeswax and wood stain from propolis which he used for his violins. His renewed enthusiasm for the craft was infectious and it was a great privilege to have him as my mentor.

Once, while I was unloading equipment from my car, he had hobbled up to the hives before putting on his veil. Although usually well behaved, on this occasion the bees took exception to his presence; they stung him mercilessly on his bare hands which were still gripping his crutches. I dragged him into some bushes, got rid of the bees and drove him home, by which time he was in shock. I was terribly concerned but all Carass was interested in was why the bees had attacked his hands. After mulling this over for several days he rang me in great excitement. He had concluded that the black grips of his crutches had triggered an instinct in the bees to attack what they thought was the snout of a bear. Bears are bees' only natural predators, and their snout is the only part of their body unprotected by fur.

That was 40 years ago. I sometimes wonder what Carass would think of the Drayton Hive. I like to think that he would approve.

Introduction

I began beekeeping in 1979, first with WBCs and then with Nationals. Soon I had 14 hives across two or three sites and was harvesting honey by the hundredweight. However, all that lifting eventually caused serious damage to my back and although I had surgery, it was four years before I could stand fully upright again. I sold the bees and gave my powered extractor to the local beekeeping association.

When I resumed, a decade later, it was with Warré hives. I no longer wanted to maximise honey production; I was relaxed about swarming and became interested in the bee's ability to defend itself against varroa. I found this a much more fulfilling and enjoyable approach to beekeeping.

However, the practice of nadiring, in which boxes are added to the bottom of the stack instead of the top as in most framed hives, threatened more back strain. But, for me, the real problem was the natural propensity for bees to attach comb to the hive walls and, sometimes, to comb in the box below. Releasing it is stressful for both the bees and beekeeper. I also tried Horizontal Top Bar Hives but, in addition to the problem of brace comb, I encountered a new one: full combs collapsed under their own weight in hot weather and the hive's design makes feeding almost impossible. There must be a better way, I decided.

Figure 1. The author with one of his Nationals in 1982. The bees, caught as a swarm the previous year, yielded 212lb of honey in a single season.

So, I started building a hive that did not require much lifting, one that used frames instead of bars but frames which could be cleared of honey without the need for expensive equipment. I had become convinced of the benefits of insulation, and I wanted to regain my shed which, for years, had been full of supers or Warré boxes.

Several prototypes later, I developed a hive which borrows some of the best features of the established designs and avoids some of the difficulties inherent in their use. What has become known as the Drayton Hive was originally just for me, but it immediately attracted the interest of other beekeepers who suffered from the same frustrations. Production of the hive on a small scale began at the Wood School of the Sylva Foundation, an environmental charity which fosters home-grown timber. However, it soon became clear that more capacity was needed, and it is now in commercial production – see www.draytonbeehive.com

A simplified approach to beekeeping

The Drayton Hive is intended for people who are content with two or three colonies of bees in hives whose appearance can enhance a garden setting more than a stack of boxes. It is intended for those who want to enjoy their hobby without feeling pressured to make regular hive inspections and whose expectations of a honey harvest are limited to what the bees can spare. This approach also results in contented bees.

Much of traditional practice focuses on interventions to prevent or control the natural impulse of colonies to raise new queens every year or two, and swarm. This different approach celebrates swarming as nature's way of re-queening and regeneration. Queens raised naturally are mated with drones from the same locality, thereby limiting the risk of infection being transmitted from elsewhere.

Traditional practice also aims to build up colonies to a size rarely seen in nature with the object of maximising honey production. As someone who has spent many years doing exactly that I have a great respect for those whose husbandry ensures a local supply of honey and of bees for pollination. There is also a satisfying intellectual challenge in close observation of hive behaviour and in devising strategies to manage it accordingly.

A more relaxed approach leaves the bees in control. Unstressed by constant manipulation, they are much more likely to be docile bees – just what is wanted in a garden hive.

Apart from the investment in hives, the traditional beekeeper's biggest expense is probably a mechanical honey extractor, necessary for dealing with frames fitted with wired foundation. However, as this little book will show, a different approach enables honey to be extracted easily and without mess, and without the need to buy a large, cumbersome machine which is out of use for most of the year.

Figure 2. Perfect for stocking an empty hive.

Essential features of the Drayton Hive

Horizontal format

In the wild, bees will occupy any cavity that suits them and are more concerned about its position and size rather than its shape. It is man's appetite for honey that, in western Europe and North America, gave rise to the vertical hive, enabling increasingly sophisticated manipulations in bee husbandry. In eastern Europe and elsewhere the culture was for keeping bees in horizontal hives, contributing to the development of designs such as the Layens, Long National and Horizontal Top Bar which are now widely used. The Drayton Hive belongs in this latter category.

The immediate advantage of the horizontal format is that the beekeeper is spared the risk of back pain caused by lifting and bending, a common reason why some are forced to abandon their craft. The disadvantage is that horizontal hives are not easily manoeuvrable so care should be taken in locating a permanent site for the Drayton Hive. However, although this little book advocates a 'relaxed' approach to beekeeping, the experienced beekeeper will find that, with a little ingenuity, many traditional practices of hive management can be adapted to the horizontal format (see, for example, *Uniting colonies*).

All-round, all-year insulation

Recent years have seen an increased emphasis on providing additional insulation for hives during the winter months but, with the inexorable rise in summer temperatures, insulation is just as important to prevent the colony overheating. Although its importance was recognised a century ago in the design of the WBC hive, the natural habitat of bees is in tree cavities that are surrounded by several inches of solid timber. Such conditions are replicated in log hives, but these pose other problems for the beekeeper.

The Drayton Hive consists of a strong inner box made from thick ply or solid timber, with a sealed air gap between it and the hive walls, usually made from cedar for its additional insulation and weather-proofing properties. This construction follows the same principles as double glazing and, although it is tempting to fill the gap with wood shavings or other natural material, few match the insulating properties of air provided it is sealed against drafts. There is a frame round the entrance and observation window to maintain this seal.

Roof insulation is provided by three hessian cushions stuffed with dust-free wood shavings or barley straw. These cushions are equivalent to 'quilts' in Warré parlance.

Figure 3 (left). Insulation cushions in place.
Figure 4 (right). Each cushion has a Velcro closure. This one is filled with dust-free wood shavings but dust-free barley straw is a good alternative.

Raised, small entrance

Other hive designs have shown that if the entrance is raised significantly above floor level, the bees are more fastidious in their housekeeping. This feature of the Drayton Hive has resulted in few dead bees or other debris left to moulder on the floor (see *Floor* and *A case study*). In addition, and reflecting the preference of bees in nature, the entrance is much smaller than in many traditional hives. It consists of a slot just 8mm high and 150mm long – big enough for workers, drones and queens to access but too narrow for mice and, by concentrating traffic into a small space, making it less likely that wasps, robber bees and wax moth etc can get in unchallenged.

Figure 5. Drayton Hives busy with bees.

There is a landing board in front of the entrance, mainly for the benefit of the beekeeper. That brief moment during which bees walk into the hive enables their pollen loads to be seen with the possibility of identifying the source of forage. Observing activity at the hive entrance is one of the many pleasures of beekeeping and one which, with practice, helps in the assessment of the colony's development and well-being (Figure 5).

Observation window

This enables a rough assessment of a colony's development to be made without disturbing it. All that is visible is the comb attached to the side bars of occupied frames but that is enough for the beekeeper to monitor the colony's expansion. The window does not show enough for the signs of imminent swarming to be recognised but it is useful in alerting the beekeeper, for example, that it may be time to remove some honey (see *Seasonal Management*). It also enables visitors to glimpse inside the hive; the sheer number of bees rarely fails to impress.

Figure 6. A Drayton Hive with its observation window open.

However, it should be used sparingly in the winter months. Not only does opening it remove the element of insulation, although briefly, but the cluster could be located out of sight, giving the false impression that the colony may have died.

Roof

The roof is pitched and overhangs the hive body to provide a measure of weather protection. It is made of cedar or other lightweight materials so can be lifted away easily; in some models it is hinged to the hive and supported open on gas struts. There are ventilation holes at each end and sufficient space under it to accommodate the insulation cushions and, when required, a half-gallon rapid feeder (see *Feeding*).

Figure 7. Air vents in the roof and the porous insulating materials under it, ensure the hive is always well ventilated.

Floor

Hive debris can be removed easily because the entire floor hinges down. As in other hives, the bees will line the entire internal cavity with a fine layer of propolis and it will require a sharp knock to release the floor. It is advisable to hold the floor in place while knocking it from above so that it does not fly open and alarm the bees. However, practice has shown that it is rarely necessary to do this more than twice a year – at the beginning and end of each season. And as noted above, the raised entrance tends to induce the bees to remove debris before it accumulates.

If required, mite drop can be monitored by inserting a card over the floor and if varroa becomes a concern, the colony can be treated according to the manufacturer's instructions.

Figure 8. A Drayton Hive with its floor in the open position.

Frames

The Drayton Hive uses 12x14 Hoffman frames (metric size 29x34cm), readily available from beekeeping suppliers. The standard hive holds 18 such frames which, in terms of comb area, is roughly equivalent to a National hive with a brood chamber and two supers. The frames are fitted 'warm way', ie across the hive and are suspended on metal runners.

Figure 9. How the frames are installed in the Drayton Hive.

An important feature of the Drayton Hive – and one of its strongest benefits – is that the frames are *not* fitted with foundation. The bees build their own comb within the frames and can decide where and when they need cells for raising workers, drones and, indeed, queens, without interference from the beekeeper. Allowing nature to take its course in this way almost always results in bees that are unstressed, safe and calm to work with, or simply to observe.

In practical terms, the absence of wired foundation means that honey can be harvested without having to use a mechanical extractor, and the method for doing so is quicker and less messy than having to uncap cells and clean the machine before and after use (see *Harvesting*). This feature is one of the greatest advantages of the Drayton Hive.

Anyone who has endured the fiddly business of fitting wired foundation will welcome this development, particularly as at least one beekeeping supplier sells Hoffman top bars with a V-profile on the underside as illustrated here. The bees will attach their comb to it and, as it grows, further support will be given by the side bars and bottom bar. A further benefit is that each frame can be assembled from the flat in less than two minutes and will last years.

Figure 10. A 12x14 Hoffman frame fitted with a V-profile on the underside of the top bar.

It is also perfectly possible to use standard Hoffman top bars if they are primed by running a bead of molten wax down the groove on the underside, or by fitting a 10cm starter strip of *unwired* foundation. The result will be the same – a series of combs separated by bee space in frames which can be lifted from the hive without difficulty. Occasionally, the combs may waver slightly and sometimes there is cross-combing but in managing the Drayton Hive, such irregularities present no significant problem.

It is perhaps worth noting here that all frames remain in the hive throughout the year; unlike most other designs, the Drayton Hive requires no extra storage space for supers etc.

Figure 11. The queen excluder.

Queen excluder

During the spring and summer months the queen excluder enables the hive body to be divided into brood and honey chambers by inserting it roughly in the middle, depending on the size of the colony. The excluder consists of the normal plastic type held rigid and vertically by a frame on either side to maintain bee space between its surface and the adjacent comb. It must also be 'bee-tight' with only a small gap between it and the floor and walls. However, there should always be sufficient frames in the brood chamber for the queen and her developing brood, a situation that can be monitored through the observation window, and the position of the excluder can be adjusted accordingly. As the colony starts to store honey, it will do so by expanding through the excluder.

There is a legitimate argument for suggesting that the excluder is unnecessary and that its use runs counter to the Drayton Hive's professed aims to simplify the practice of beekeeping. Without the excluder, the cluster is likely to drift towards the middle of the hive and honey may be stored at both ends. That is not a significant problem although the cover cloth will have to be removed from the brood chamber in order to lift frames from the front if required for extraction, causing more disturbance than if they are withdrawn only from the back. Leaving stores at both ends of the hive for winter use risks isolation starvation.

Division board

As with most other hives, it is expedient to reduce the space occupied by the cluster during the winter months in order to conserve heat. In the Drayton Hive this is done by replacing the excluder with a division board, made to the same dimensions as the queen excluder. The space behind it is used for storing frames which should be empty of comb at that time of year (see *Seasonal Management*).

Figure 12. An occupied Drayton Hive opened to show the division board.

Cover cloth

It is necessary to fit a cover cloth over the frames to separate them from the insulation cushions. This replaces the crown board of most conventional hives and, following traditional Warré practice, it is made from stiffened hessian

– stiffened because natural hessian frays. It will not prevent a few bees leaking into the roof space but that is not a problem; in the Drayton Hive, bees are free to roam. The bees will propolise the cover cloth down but, because hessian is porous, they will also strip propolis away when they need to increase ventilation and prevent the hive from overheating. The use of porous materials in the cover cloth also enables the insulation cushions to absorb excess condensation which can be a killer, particularly in winter.

A hole, of about 10x10cm, is cut towards the front of the cover cloth, normally covered by a larger piece of stiffened hessian to prevent bee leakage. This hole enables bees to access a half-gallon rapid feeder or a block of fondant when required (see *Feeding*).

Figure 13. The feeding hole through the stiffened hessian cover cloth with the additional piece of stiffened hessian to cover it when not in use.

Stiffened hessian cover cloths only last two or three seasons because, eventually, the bees will start chewing them. However, hessian, both natural and stiffened is very cheap and the cost of replacing them is hardly significant. Stiffened hessian is commonly used in model-making and in upholstery. It is readily available from fabric shops and online; it can also be made by painting both sides of natural hessian with a thin flour paste. The cloth will shrink while drying so should be cut to size afterwards.

Some Warré users are now replacing stiffened hessian by fly net with the advantage that the bees are unlikely to chew it, and it should last years. This material should offer the same benefits in the Drayton Hive.

Feeding

Liquid feed can be offered to the bees in a Drayton Hive simply be removing the front insulation cushion, peeling away the stiffened hessian over the hole in the cover cloth, and installing a half-gallon feeder over it. Pour in syrup of the correct consistency (stiff in the autumn, thinner in the spring – see *Seasonal Management*) and replace the roof. Exactly the same procedure should be followed for installing a block of fondant.

Figure 14. Feeding time!

A better alternative, and one which certainly aids bee health, is to ensure that the colony always has sufficient honey in the hive so that feeding is unnecessary. In preparation for winter, this means leaving a minimum of three frames of fully capped honey in the brood chamber, together with one or two half-full frames in which the bees can store their late season forage (see *Seasonal Management*). This is considerably less than the amount advocated in traditional beekeeping practice but, over several seasons, it has been found to be sufficient. Good insulation decreases the consumption of winter stores which probably explains the survival rates in the Drayton Hive. However, as a precaution, it is expedient to offer fondant in the following spring,

Harvesting

Honey

In the spring the colony will expand through the queen excluder and begin storing honey in the frames behind it. Naturally, they will fill the frames closest to the excluder first, and when these frames have been at least 70% filled with capped honey, they are ready for extraction. A rough judgement on development can be made through the observation window without having to open the hive. A conclusive assessment can be made by peeling back the cover cloth from the back of the hive but no further than the excluder so as not to disturb the brood chamber. The frames at the front of the honey chamber can then be lifted out. This is done most easily by first removing a frame at the back, to allow more space in which to operate.

Prototypes of the Drayton Hive used a bee escape to clear frames of bees, but this has proved to be unnecessary. Most of the bees clinging to the surface of the comb can be removed by gently shaking the frame over those remaining in the honey chamber and by brushing off the stragglers. However, frames being cleared in this way should always be held in the vertical plane to avoid the comb dropping out under its own weight of honey.

The frame of honey, now cleared of bees, is then held over a suitable container (typically a polythene box at least 30x40cm with a lid) and the comb cut from it with a

Figure 15. Removing a full frame of sealed honey.

Figure 16. Cutting the comb from the frame. The wet frame is then returned to the hive.

serrated knife, so that the comb falls into the box. The wet frame should then be returned to the honey chamber and the box closed up so as not to attract flying bees. It is a very quick and easy operation, and several frames of honey can be loaded into the box in this way.

All that is required for extraction is a honey tank fitted with a strainer and tap. In a bee-proof room, combs are manually crushed into the strainer. A complete frame of honey can be squeezed out in about two minutes, yielding up to 5lb of honey. It is then left to drain through the strainer overnight before bottling. Two or three frames can be extracted at a time, a process that can be repeated little and often throughout the season. There is an advantage in crushing the comb into a domestic colander placed on the strainer; this prevents the strainer becoming blocked and leaves a residue of soft, new wax that has been partially drained of honey.

Figure 17 (left). Honey tank ready for action.
Figure 18 (right). A domestic colander can be used to collect the crushed wax to speed up the flow of honey through the strainer.

If it is done methodically, this process can be completed quickly and without mess. Work surfaces should be covered in newspaper and disposable gloves used for the manual crushing, and both are then thrown away after use.

Beeswax

The Drayton Hive offers two sources of wax. The first is a by-product of honey extraction, as above. The second is from the removal of empty comb when consolidating the brood chamber at the end of the season (see *Seasonal Management*). To obtain pure, clean wax it has to be rendered. This can be done in a solar wax extractor, either home-made or available from most beekeeping suppliers, or by soaking the crushed comb in water for a few days, then boiling it in a weighted muslin bag. The wax will rise to the surface of the water, leaving any impurities in the bag. When the water has cooled, the clean wax will form a solid disc floating on its surface, which can be easily removed.

Figure 19. Crushed wax ready for rendering.

The resulting wax can then be used in making candles, furniture polish and a range of cosmetics. There is also a market for it among beekeeping suppliers who melt it down for reselling as foundation. Weight-for-weight, beeswax usually has a higher value than honey.

Hive hygiene

The system of hive management advocated here results in the gradual movement of frames from the back of the hive to the front. During the process none of the comb is likely to have been in the hive for longer than three seasons before it is cut out (see *Seasonal Management*) which lessens the likelihood of it harbouring the infections which can occur when old comb is re-used.

The Drayton Hive is also less likely to attract wax moth because, when adult moths are active, bees are working on the comb either in raising brood or storing honey; otherwise, the frames are empty. As with all other hives, though, bees in the Drayton Hive are at risk from varroa, foul brood and other diseases.

There is no doubt that this process of constant wax renewal is a drain on the colony's resources which might otherwise be used in honey production. However, it is also likely that it restrains the swarming impulse provided the hive is managed so as to avoid congestion.

Seasonal management

Winter

During the dark, cold months of January and February the colony, now greatly reduced in size, will be tightly clustered round the queen. She will probably not be laying, and the cluster will be located towards the back of the brood chamber, gradually eating its way through the stores towards the division board. Because the frames are set 'warm way' the comb in those at the front of the hive will act as a baffle against wind and rain that might otherwise get through the entrance. On warm, sunny days it is always encouraging to see bees making short defecation flights, collecting water or, even, pollen from hazel, hornbeam and early bulbs, but there is nothing for the beekeeper to do except look forward to the spring.

Spring

The beginning of spring is an anxious time of year because this is when the colony is at its most vulnerable. If the bees have consumed their winter stores and the spring is delayed by wet or cold weather, starvation may result. It is expedient, therefore, to offer fondant (see *Feeding*), although experience has shown that bees in the Drayton Hive seem to fare better than in hives without the same level of insulation.

The queen will have started to lay and by April the colony should be expanding rapidly. On a warm, still day when the bees are flying well, the hive should be opened, and the division board replaced by the queen excluder. Those who want to satisfy themselves on the wellbeing of the colony could undertake a quick inspection of frames in the brood chamber. This is also the time of year for opening the floor and disposing of any dead bees and other debris that may have accumulated through the winter. However, these are disruptive activities, and the hive should not be left open for longer than is necessary.

Towards the end of April and into May the colony should be expanding through the queen excluder, building comb in the empty frames of the honey chamber, and beginning to fill them with nectar. The situation should be monitored through the observation window and if there is a strong honey flow, causing the frames at the front of the honey chamber to fill quickly, they should be moved further back and replaced by emptier frames. Failure to do so risks congestion in the hive which might stimulate swarming.

By the end of May there may be two or three frames full of sealed honey which can be extracted (see *Harvesting*), further reducing the possibility of congestion. If the bees are likely to be foraging on oil seed rape, frames which have become at least 70% full of sealed honey should be extracted as soon as possible to avoid them solidifying in the hive.

Summer

The colony's expansion will continue through June unless a lack of local forage results in the 'June gap'. Its development should continue to be monitored through the observation window and frames re-arranged in the honey chamber to avoid congestion as necessary, and more honey could be extracted. By the middle of July, the colony will reach the peak of its expansion; more honey can be extracted but it is expedient to discontinue doing so into August so that the bees can fill up two or three frames for their own winter stores.

Wasps will be active but, because of the small entrance to the Drayton Hive, the colony should be able to defend itself. However, if wasps are seen to be getting in, the entrance can be reduced in size with a block of wood, temporarily held in place with drawing pins or Blu Tac.

Autumn

It may feel like the height of summer but, by the end of August, the colony will have contracted as it prepares for winter, and the beekeeper should help in these preparations. In the Drayton Hive, this involves removing any frames from the front of the brood chamber that are completely free of brood, honey and pollen, and moving the rest of the frames forward. The lack of stores in the front of the hive eliminates the risk of the colony suffering from isolation starvation during the winter. This is also the time to release the floor and clear it of any debris that has accumulated during the season.

The queen excluder should be replaced by the division board and any frames from the honey chamber that still contain honey should be placed in front of it as stores to see the bees through the winter. The occupied part of the hive is thus consolidated to about half its volume with, ideally, the following arrangement of frames, starting from the front: those from the brood chamber still covered in bees and including the queen and brood; two or three partially full frames from the honey chamber; three full frames of honey, sealed or unsealed. Those applying autumn treatments for varroa should do so now, making sure they follow the manufacturer's instructions.

All comb in the frames behind the division board should be cut out, leaving just a line of wax under the top bar as a starter strip for the bees when the frames come back into use the following spring. It is probably a good idea to remove surplus propolis from the frames at the same time.

If it has not been possible to provide the colony with three full frames of winter stores, it will need feeding with thick syrup (see *Feeding*) at a rate of a full half-gallon feeder per frame. The two or three partially full frames remain available for the bees in which to store autumn forage, a process which can continue into November. The cluster will be at the front of the hive and, as the winter season progresses, it will eat its way through the stores towards the centre of the hive, leaving the empty comb in the frames at the front to act as a baffle against wind and rain.

Precautions should continue to be taken against wasps. If the hive is in an exposed site, it may be expedient to put a strap round it to prevent the roof blowing off.

And so the cycle repeats itself.

Some personal reflections on beekeeping practice

The Drayton Hive is a hybrid and so is the system of management that evolved in its development. Along the way I discovered a strong body of opinion that had reached better-informed beliefs than my own questioning of traditional beekeeping practice. There are now thousands of 'natural' beekeepers, of which I claim to be one, but I am not a hard-liner and I continue to compromise in my own practice.

Figure 20. A Drayton Hive, recently occupied by a swarm, showing how the bees build comb on foundationless frames.

For example, I am not prepared to allow a weak colony to face the winter without some help. That means feeding and, perhaps, uniting with another weak colony (see *Uniting colonies*). I would, however, be the first to acknowledge that there is nothing 'natural' about this procedure, and it risks spreading infection between colonies.

All living creatures have a capacity to tolerate disease unless it becomes overwhelming. It is reasonable to assume, therefore, that there is a measure of varroa in most untreated colonies, including the bees in my own Drayton Hives. Only once have I suspected varroa as the cause of death in my colonies, and once they were badly affected by deformed wing virus. Varroa originated in the Far East where it is now in decline because the natural populations of bees there have evolved their own mechanisms for combatting it. Those same mechanisms are beginning to take effect in Europe, and I support the growing body of opinion which believes that chemical treatments and other interventions serve only to delay the opportunity for bees to build up their own resistance to the disease.

I have always stocked my hives with swarms and am one of the local swarm collectors. For me, catching a swarm and hiving it successfully, or passing it on to someone else in need, is one of beekeeping's greatest pleasures. I always make sure that those who call me, often because a swarm has settled in their garden, watch the process from a safe distance and are, later, rewarded with a pot of honey. Rarely have they seen so many bees at close quarters and are fascinated; if they had the privilege of seeing the swarm in flight, they are awed. Installing a swarm into an empty Drayton Hive is best done by shaking it into the hive body rather than by 'walking it in' because of the relatively small entrance which could become congested with bees.

Traditional beekeeping practice involves close inspection of every frame in the brood chamber every week between early April and late July with the principal aim of spotting the early signs of swarming, and of acting accordingly. When I was a 'traditionalist' I challenged myself to find the queen every time I opened a hive; it is a knack which comes with practice, and I became quite good at it. I also did my own queen-rearing. Success with the Drayton Hive does not depend on these skills but there is nothing more reassuring to the beekeeper than seeing a queen going quietly about her business. However, those skills require sharp eyesight and a steady hand which, with a strong back, are features which the beekeeper cannot always rely on as the years advance.

Uniting colonies

As well as my Drayton Hives, I have a deep-bodied National which I use for housing swarms if I have one to spare. This is particularly useful for uniting with one of the other colonies to ensure it is strong enough to see it through the winter. There are risks, of course, of which one is that both queens die in the conflict between them, and it may be too late in the season for the united colonies to raise new queen cells.

The traditional method for uniting colonies is to arrange for the hives to be placed close to each other for a week or two so that the bees in both are established in the same location. One hive body is then placed on top of the other, separated by a piece of newspaper with a few perforations in it so that the scent of both hives mingles. The workers in both hives chew through the paper and, in doing so, become accustomed to working together, but the queens fight. Usually, the fittest and youngest wins and inherits the united colony.

Exactly the same process can be undertaken in the Drayton Hive except, of course, that the newspaper separating the colonies must be supported vertically. I use a board I contrived for one of the prototype hives; it is of the same dimensions as the division board, with newspaper taped around the sides and bottom, and over a whole in the middle.

Figure 21. A simple method for uniting two colonies.

This is inserted behind the occupied frames in the Drayton Hive and those occupied by the swarm in the National are placed behind it. I prefer to do this early in the morning, before the bees are flying, so that the bees from the National can re-locate themselves as they emerge from the Drayton. It works.

Figure 22. Three prototype Drayton Hives and a production model from Sylva with the deep-bodied National used as temporary housing for swarms.

A case study

During the heatwave of 2022, one of my prototype Draytons became bearded with bees clustered all over the front of the hive. This often happens in particularly hot weather with normal, uninsulated hives but shouldn't happen with a Drayton. On inspection I found that the cause was not simply the heat; a joint in one of the brood frames had come apart and the weight of honey had caused it to collapse, bringing down comb from the adjacent frames as well. The floor was covered in slabs of brood and honeycomb, honey to a depth of 1cm, and hundreds of struggling bees. I attempted to prop up the broken combs between the frames, but some simply had to be removed. It was a couple of days before I was able to return with bits of kit that I hoped would help in clearing the mess from a hive that was, still, fully occupied. In the meantime, though, the bees had returned to the hive and had done their own housekeeping. Only a few dead bees remained from the chaos and the floor was hardly sticky. It was a reminder of what remarkable creatures honeybees are.

I assumed, however, that the queen had not survived the experience, so I decided to unite the colony with a swarm that I had already hived in the National. I was planning to do this anyway so, fortunately, it was already in position. The apiary, consisting of three prototype Draytons, a Sylva production model, and the National is illustrated here. The operation went smoothly, and within 24 hours, the Drayton had a good, strong colony, hopefully headed by a satisfactory queen.

Several lessons could be learned from this incident, including is the importance of discarding old and wonky frames.

Figure 23. A Drayton Hive built by the Sylva Foundation. The frame and legs are oak and the cladding is cedar.

Who is the Drayton Hive for?

The short answer to this question is – people like me. I have served my apprenticeship and taken exams so I claim quite an understanding of bee behaviour and how it can be manipulated to serve man's needs. Such knowledge and experience add greatly to the fulfilment to be gained from keeping bees. I therefore see the primary market for the Drayton Hive as being among experienced beekeepers who want a hive that is less demanding in its management. Those who respond to the various magazine articles about the Drayton Hive, and to www.draytonbeehive.com certainly fall within this group.

A second, unexpected market has emerged among those who are interested in using the Drayton Hive as their introduction to beekeeping. Personally, I think it is probably best to get started with one of the standard hives to give the beginner a better insight into the craft than is possible with the Drayton Hive. However, as always, a good mentor can make up for a lot of inexperience.

Another factor for consideration is the cost. The Drayton Hive, with its double walls and other features, is time-consuming in its construction and this has to be reflected in its price. The Drayton Hive costs about three times as much as a basic National package. Against that, there is no need to buy foundation or a mechanical extractor, and no space is required for storing supers, for example. But for those who have the workshop and necessary skills, some basic plans are included at the end of this book.

Further information

This is a book about a particular type of beehive and, at the time of writing, it is the only book on the Drayton Hive there is. It is not a book about beekeeping.

However, the Drayton Hive's development is based on two traditional types, the WBC and the National and two which have increased in popularity in recent years, the Warré and the Horizontal Top Bar. There is an extensive literature on traditional hives and their management, from which I recommend:

Hooper, Ted (2010) *Guide to Bees & Honey, 5th Edition* Northern Bee Books

Waring, Claire & Waring, Adrian (2011) *Haynes Bee manual: the complete step-by-step guide to keeping bees* Haynes Publishing

Wedmore, E.B. (1989) *A Manual of Beekeeping, 3rd Edition* Northern Bee Books

This last was the 'bible' for an earlier generation of beekeepers and it is still my reference of choice when I need detailed information on a particular problem. An early edition is part of a collection of beekeeping books, some of them going back to the 19th century, bequeathed to me by my mentor, Carass (see *Preface*).

Natural (or non-invasive) beekeeping and its various hives has its own literature and I recommend:

Chandler, Philip (2015) *The Barefoot Beekeeper* lulu.com

Chandler, Philip (2015) *Managing the Top Bar Hive* lulu.com

Heaf, David (2013) *Natural Beekeeping with the Warré Hive: a manual* Northern Bee Books

Heaf, David (2021) *Treatment-free Beekeeping* Northern Bee Books

Seeley, Thomas (2019) *The Lives of Bees: the untold story of the honey bee in the wild* Princeton University Press

Storch, H. (2012) *At the Hive Entrance* European Apicultural Editions

Storch would be particularly useful to those interested in bee behaviour; it is a classic work which gives a good indication of what is going on inside the hive without having to open it.

Northern Bee Books probably has the most comprehensive stock of new, second hand and antiquarian books on every aspect of bee husbandry and scientific study. See www.northernbeebooks.co.uk

I subscribe to the UK's primary magazines, *BBKA News* and *BeeCraft*. The former used to be entrenched in the 'traditionalist' camp, but a new, enlightened editorial policy now includes articles on aspects of natural beekeeping as well. Both magazines have featured the Drayton Hive.

There is, of course, an endless resource of information online, including YouTube clips, often from excitable Americans demonstrating techniques while wearing a T-shirt and shorts when protective gear might be advisable. Perhaps more useful are the instructional videos from Black Mountain Honey and Simon the Beekeeper, among others. Useful websites include:

www.bbka.org.uk

www.bee-friendly.co.uk

www.dave-cushman.net

www.draytonbeehive.com

www.naturalbeekeepingtrust.org

www.oxnatbees.wordpress.com

Nothing beats communication with other beekeepers, so I strongly recommend active membership of a local beekeeping association. The BBKA (British Beekeepers Association) maintains a national database of swarm collectors – see www.bbka.org.uk/find-a-local-swarm-collector

Home construction

One of the manufacturers of the Drayton Hive is the Wood School of the Sylva Foundation, an environmental charity based in Oxfordshire. It supports creativity and craftsmanship using home-grown timber, and customers for its products include Oxford University and the National Trust. Copyright to the plans below belongs to the Sylva Foundation and are made available to beekeepers on the firm understanding that their hives are for their own use and not for sale. Those interested in licensing the commercial production of the Drayton Hive are invited to discuss the matter through www.draytonbeehive.com.

Figure 24. The production team at the Sylva Foundation.

It should also be recognised that the insulating features of the Drayton Hive require real woodworking skills, precision and experience, as well as a well-equipped workshop for it to be built satisfactorily. The Sylva Foundation cannot undertake to answer queries or advise on its construction.

Drayton Hive – Carcase A

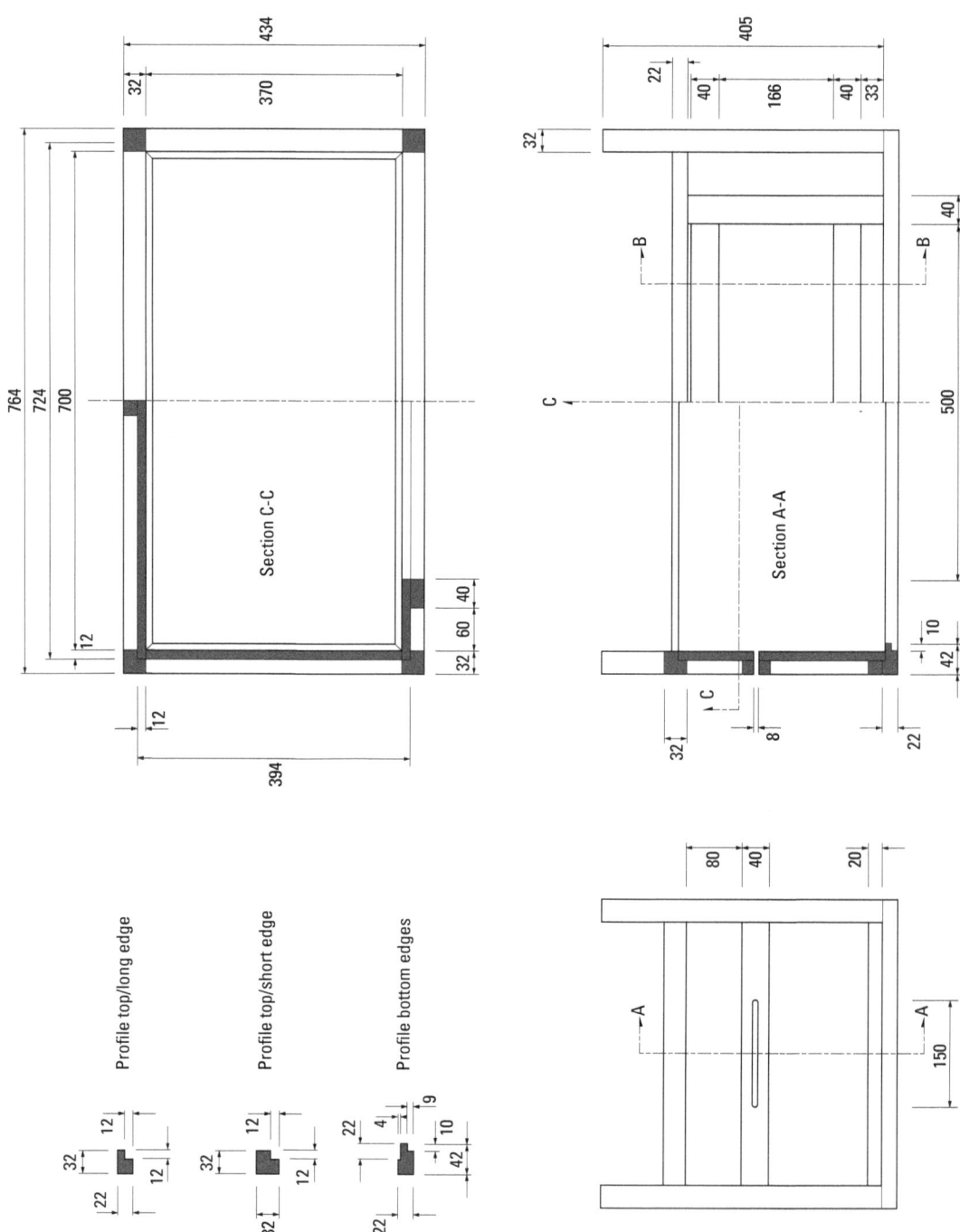

Plan 1(a). Internal carcase made from 12mm exterior grade plywood. Posts and rails made from oak.

Drayton Hive – Carcase B

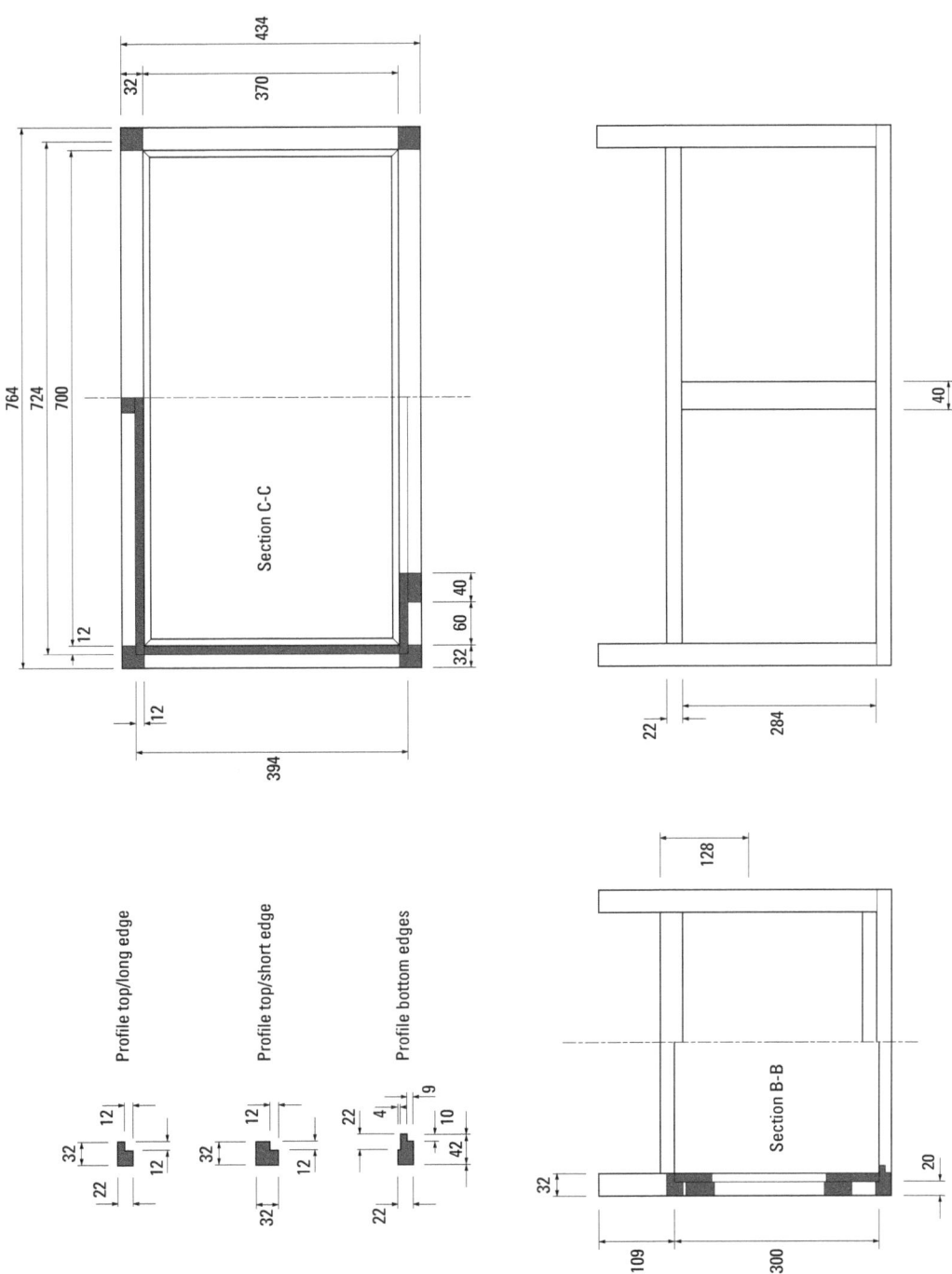

Plan 1(b). Internal carcase made from 12mm exterior grade plywood. Posts and rails made from oak.

Drayton Hive – Cladding

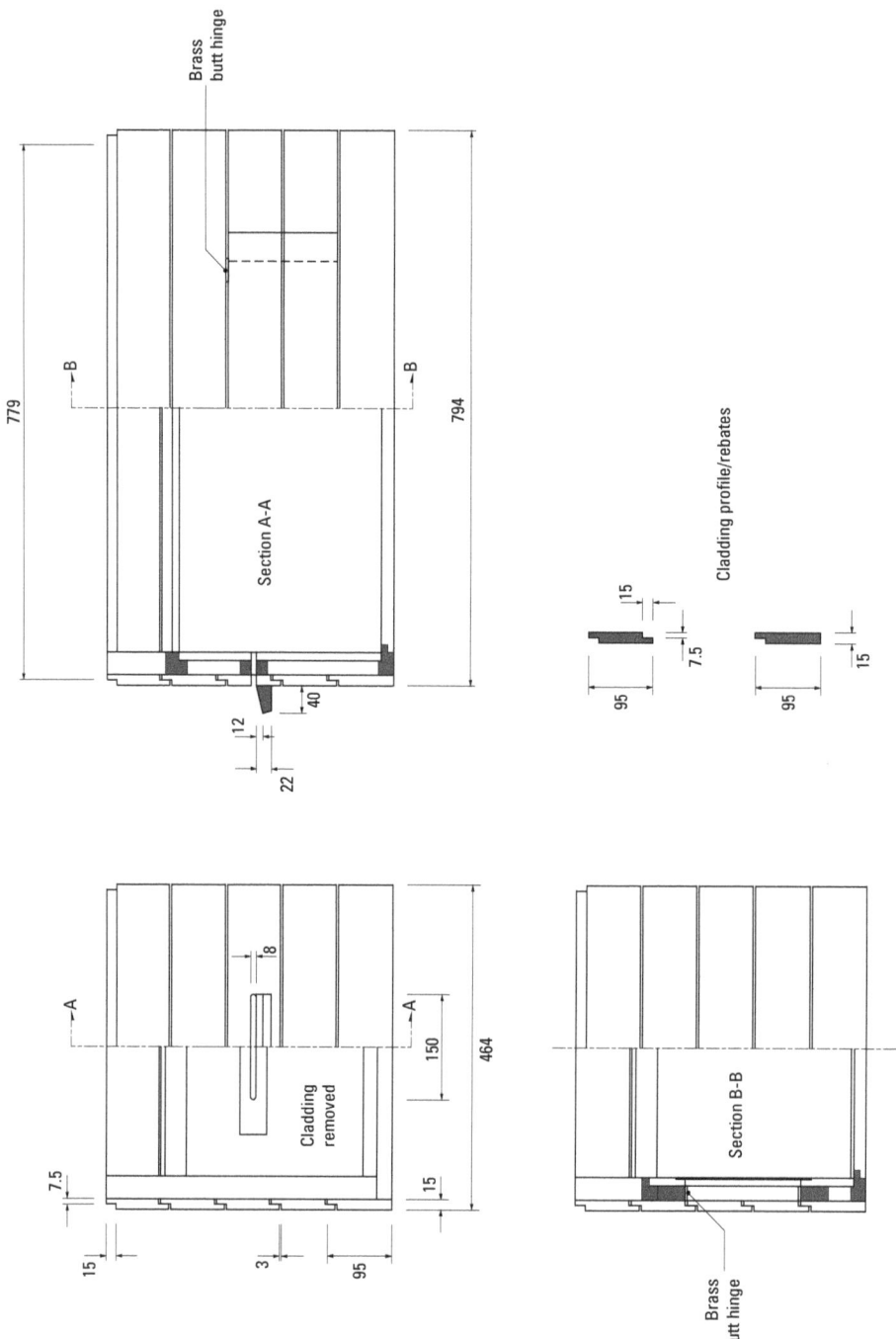

Plan 2. Cladding in western red cedar, fixed with stainless steel screws. The observation window is rebated internally with 2mm Perspex, bonded in place. The floor is secured with two brass butt hinges.

Drayton Hive – Lid

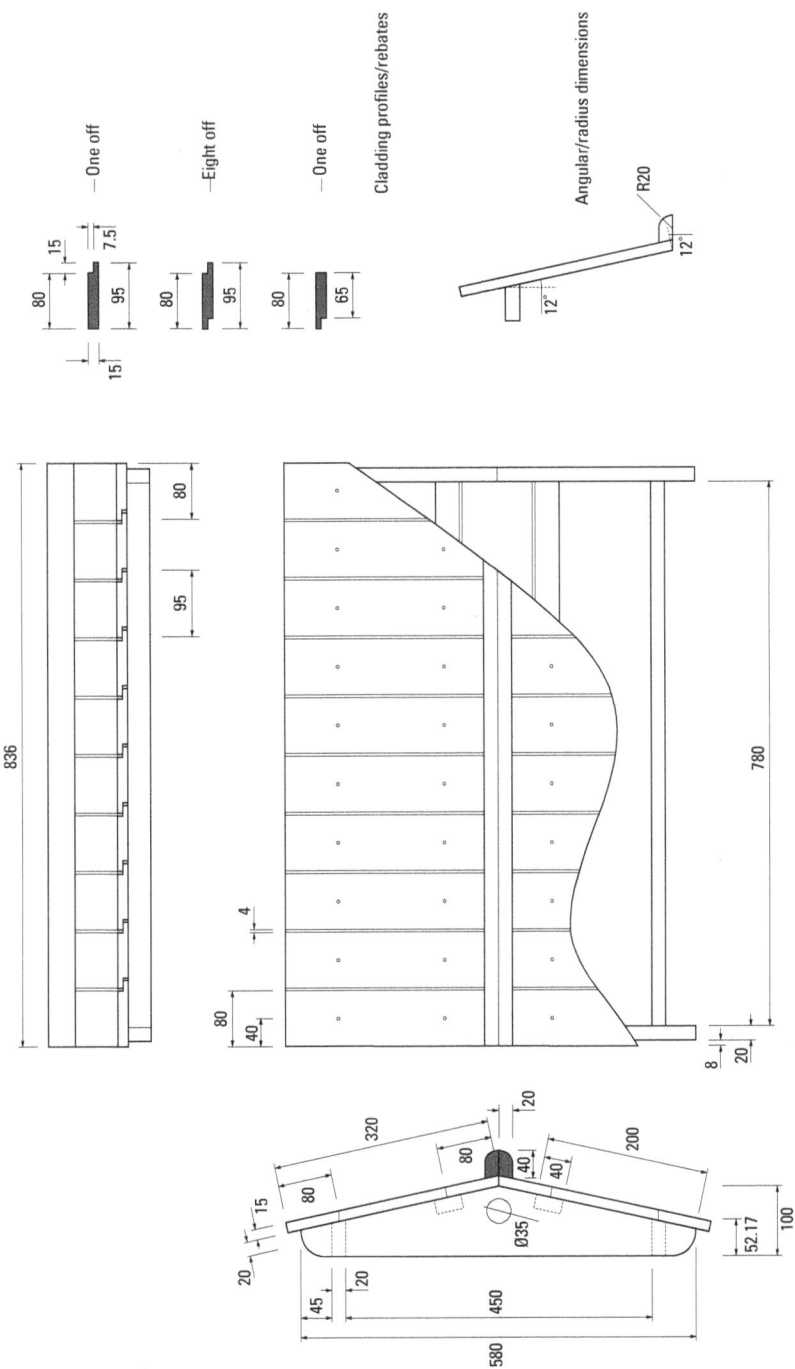

Plan 3. The roof, clad in western red cedar with a bead of silicon sealant along the joints on the underside. The ventilation holes are covered in fine steel mesh.

Drayton Hive – Underframe

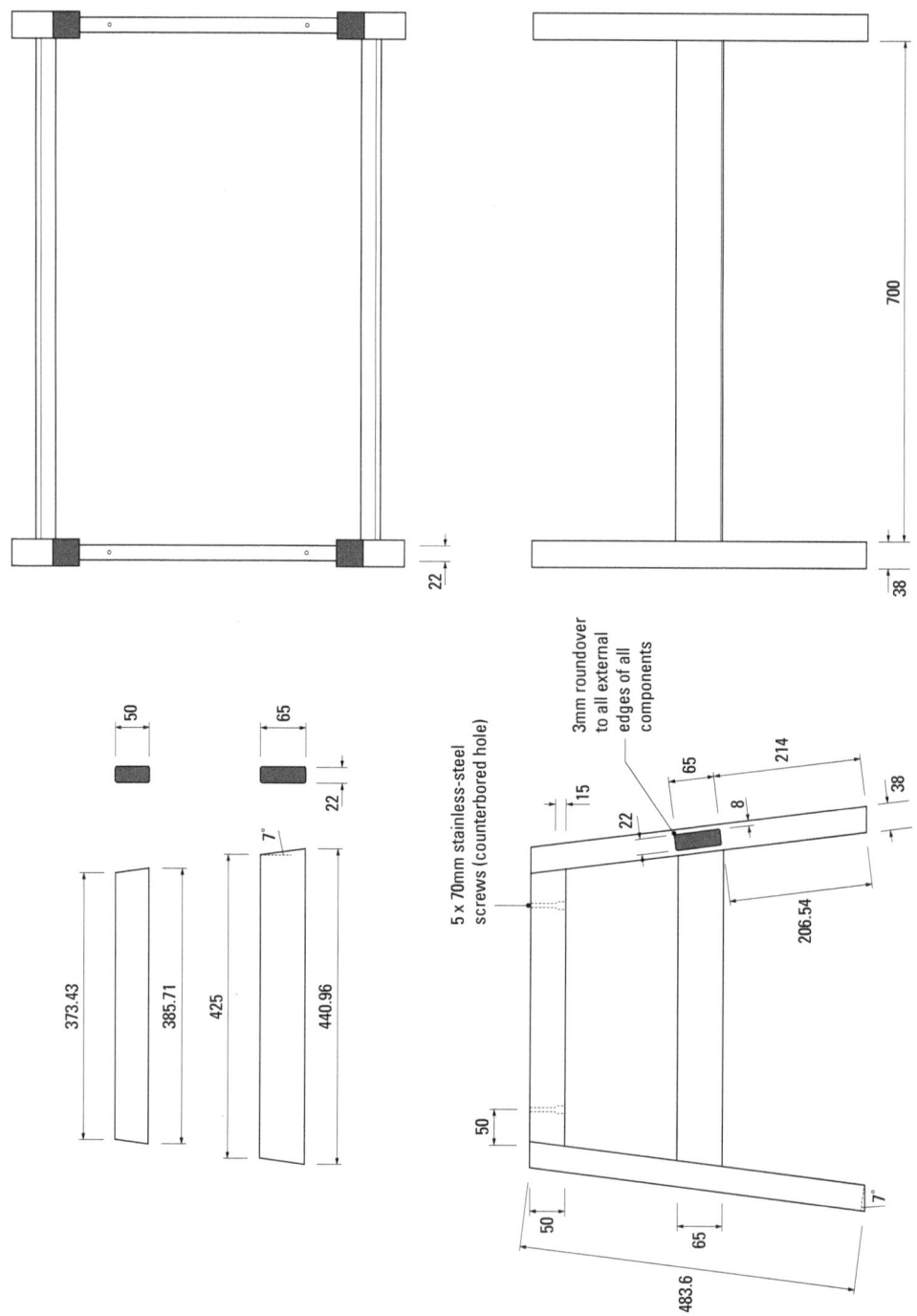

Plan 4. The underframe, constructed in oak and assembled with waterproof adhesive.

48

Drayton Hive – Internals

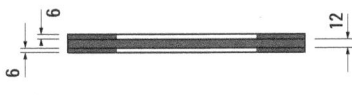

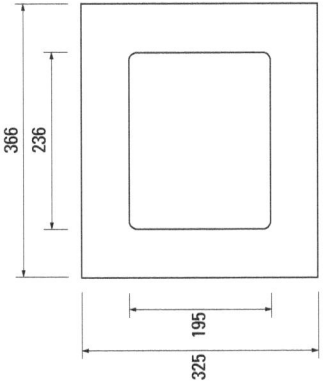

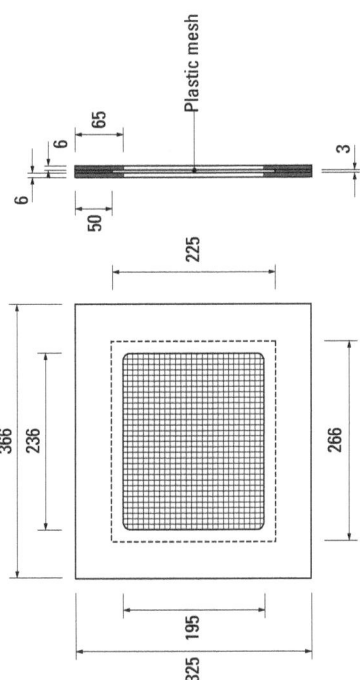

Plan 5. Left - the optional queen excluder, made with 6mm exterior grade plywood framing standard plastic queen excluder. Right – the division board made from 6mm exterior grade ply framing 12mm exterior grade plywood inner. Both assembled with waterproof adhesive.